The Universe

By Forester de Santos

The Universe / Forester de Santos

© 2019, Forester de Santos

ISBN: 9781097789801

All Rights Reserved

No part of this book may be copied, sold or distributed, in either printed or electronic format, without the written permission of Forester de Santos.

Kindle Edition

Something about the Universe

The universe or the vacuum of space really is a neutral point or point zero or a starting point and the universe becomes positive or turns into a one or a beginning when matter enters the universe or the vacuum of space from the outside of the universe or the vacuum of space.

The universe becomes negative when the largest of stars implode into giants black holes. It is through the giant black holes that the universe or that the vacuum of space is cleared or cleaned from any matter or stars or planets which are still left in the universe or in the vacuum of space and thus returning the universe or the vacuum of space into a neutral point when all of the giants holes are dispersed or turned off because of lack of energy or lack matter in the universe or in the vacuum of space.

Tags: universe, vacuum of space, matter, energy, light, black hole, positive, negative, neutral, beginning, end, and implosion, planets

About this Author Beloved

All writers must come sooner or later to the things that they want to truly write about and most come to write where the easy money is and most writers make a very big killing in becoming very rich and very famous by writing fiction and good for them!

But the great question is how much fantasy for the human race since the human race has being living in a fantasy since the race entered into consciousness or began to think and invent with words?

Well, that is a great or tall question that Forester de Santos as a writer has truly asked!

And so he has truly chosen to walk or to actually write on the road less taken or less written about!

And so he began to research and write about immortality and as the same goes, one becomes what one thinks or writes or even reads about the most!

Prologue

The universe or the vacuum of space is a neutral point or is a point zero which is converted into a positive point or into a one or more when light or matter or energy enters into the universe or into the vacuum of space from outside the universe.

The universe or the vacuum of space also is converted into a negative point when the grand majority of light turns off or the majority of matter no longer has energy or fuel…

Acknowledgement

I would like very much to give thanks to my youngest son for suggesting to me to write this short work about the Universe.

Thanks Fred!

Dedication

The Universe is dedicated to all of those that seek the truth so that the she frees them and they thus truly become for much more for they truly becoming for the very truth herself through a higher or through a taller mental consciousness or life renewed…

Table of Contents

Something about the Universe..3
About this Author Beloved..4
Prologue..5
Acknowledgement...6
Dedication...7

Introduction..10

Chapter One..12
What is space? 12

Chapter Two..25
As it can really be seen..25

Chapter Three...28
Dimensions of time..28

Chapter Four...34
The Stars and the Black Holes...34

Chapter Five..40
The Universe and the Number..40

Conclusion..47
Further Notes...50
The (0+-) Factor...53

Who am I really? 55

Introduction

What makes the universe?"

Or "what really makes possible the universe?"

That is to really say, what is the universe made up of?

In other words, what are the real things that actually make but possible the universe?

What are the recognized physical things that really make possible the universe?

And how these recognized physical or recognized solid things really make the universe round?

The universe is made up of space, darkness, and coldness.

The universe is also made up of matter, light, and heat.

The universe is also round or sphere-like and the universe really is also moving or expanding throughout the vacuum of space or moving throughout the sphere or throughout the round and invisible shield.

In short, the universe is simply made up of space, darkness, and coldness.

The universe or physical creation also is made up of solid matter, light, and heat. Motion or movement is also a major part of the universe.

And all of these things are really interacting and are expanding inside a so-called sphere or in a round and invisible but real field.

Chapter One

[(((((((-) (0) (+))))))))]

What is space?

What, then, is space?

Space is the absence of a matter, light or energy.

Space is the absence of a physical or of a solid body or the physical absence of a solid or of a physical identity.

Space also has no mass or space has no density or weight.

Space is, therefore, non-existence or nothing!

Space is also a vacuum or emptiness.

Space, is therefore, the absence of a physical reality.

That is to really say, space is nothing, space is empty and space is lacking.

Space is measured, however, not by the amount of space or the amount of emptiness but by the physical size of the actual or the real thing or the physical or the solid matter, light or energy or even heat that is going to actually fit or actually occupy that space or that emptiness.

So in reality, space in the physical sense simply does not exist or space is non-existence!

[((((((((-) (0) (+))))))))]

What is darkness?

Darkness is also the physical absence or the lack of light or Darkness is also the physical absence of energy.

Darkness also has no mass or Darkness has no density.

And what is cold or coldness?

Cold or coldness is also the physical absence or the lack of heat.

Cold or coldness is the absence or is the lack energy or heat.

Cold or coldness also has no mass or cold or coldness really has no density.

[(((((((-) (1) (+)))))))]

What is matter?

Matter is that which occupies space and matter is also that which has mass or volume.

Matter is that which is physical or is solid.

Matter is that that has mass and matter is also that which has density.

Matter is also physically affected by motion or by physical movement or by time.

Matter also physically exists or matter simply but physically is.

Matter, is therefore, physical reality.

Matter simply is physical knowledge of existence or actual reality or existence!

That is to really say, matter is also a product of physical reality or physical creation.

In fact, if no matter, then no physical reality, no universe or no physical creation!

[((((((((-) (1) (+))))))))]

What is light?

Light is that which occupies space. Light also has mass and light also has density.

Light also physically exists or light simply but is.

Also, light is a physical product of physical reality or of physical creation.

The Universe / Forester de Santos

[(((((((-) (1) (+))))))))]

What is heat?

Heat is that which occupies space.

Heat also has mass and heat also has density.

Heat also physically exists or heat simply but is.

Also, heat is a physical product of physical reality or of physical creation.

Thus, darkness is to space as light is to energy or to matter!

That is to say, darkness is to space or to cold or to coldness as matter is to energy or to light!

In short, space is to vacuum or to emptiness, to darkness and to coldness as matter is to motion or to expansion, to energy and to light.

Also, space is to nothing or space is to nothingness as matter is to something or is to physical reality.

In reality, therefore, space is to complete emptiness or space is to unseen reality as matter is to physical fullness or to physical reality.

Now let's simply say once again that space, darkness and coldness are simply but the very same thing.

And let's say also once again that matter, light and heat or energy are simply but the very same thing.

Now, there are two very simple opposites to work with or to work in: Nothing and something, or space and matter, or Non-existence and physical Existence, or Negative and Positive, or Black and White.

Or simply, that which is and that which is but not!

Space, coldness or darkness and matter, energy or light actually or really form or create the universe or physical creation, which is round or sphere-like, where space, coldness or darkness and matter, energy or light simply unite, physically join or simply intersect or physically meet or even divide.

Now, what is the solid or the physical thing with mass or with density that makes but a simple sphere or even a round shield?

What makes but a field or a sphere or even a simple shield?

What two figures or what two objects or what two things make a simple sphere or a simple circle where the two things physically unite, physically meet, physically interact or physically join or even divide?

What really makes but a sphere or shield?

What really makes a round field?

Or what is the very thing that makes a field?

What about two magnets?

What about two magnets in the form of cubes?

divide, many magnetic circles or many magnetic spheres are physically formed!

But the magnetic circles or the magnetic rings or the magnetic fields or magnetic shields must be sphere-like, for they are really the product of two magnetic cubes.

In reality, countless numbers of magnetic circles or countless numbers of magnetic spheres are formed for the Cartesian coordinate system goes on forever and ever.

Also, the magnetic circles or the magnetic spheres or the magnetic rings increase in size as real or as whole numbers increase in size by one.

That is to simply say, the size of the first magnetic sphere or magnetic ring is one or is number one.

The second magnetic sphere or magnetic ring is two or twice the size as the first magnetic sphere or magnetic ring.

And the third magnetic sphere or magnetic ring is three times the size of the first magnetic sphere or the first magnetic ring, and so on and so on.

Now, inside one of those magnetic spheres or inside one of those magnetic circles really is the universe!

When the two magnets, in this case the space cube and the matter cube or the light cube, physically attract one another through their magnetic fields, the two magnetic cubes create many magnetic spheres, [(((((((-) (0) (+))))))))].

Inside one of those magnetic spheres is the universe, [(((((((-) (U) (+))))))))].

But what, therefore, makes possible the magnetic cube?

What makes possible the cube?

That is to say, how does the cube gets its unique shape?

Since the infinite magnetic cube really is outside of space or is outside any vacuum of space or is outside any gravity or is outside any compression, there really is no compression or is no vacuum to physically act or to physically interact upon the infinite magnetic cube, thus really allowing the infinite magnetic cube to expand throughout and, therefore, making or taking its unique cube shape.

Now there are two infinite magnetic cubes and, where they physically meet, physically interact or physically intersect, countless magnetic spheres or countless magnetic circles are formed; thus, the physical universe.

One of those magnetic spheres or one of those magnetic circles is the physical universe.

Actually, every magnetic sphere or every magnetic circle is really a separate universe or a separate dimension of time!

Both infinite magnetic cubes physically attract one another or one cube physically attracts the other cube with the very same magnetic force or magnetic attraction.

That is to say, one magnetic cube attracts or physically pulls the other magnetic cube with equal magnetic force or equal magnetic attraction.

The infinite matter, the infinite light or the infinite energy cube physically attracts the infinite space, the infinite coldness or the infinite darkness cube, but the infinite space cube releases nothing for infinite space, being empty, has nothing to release.

The infinite space, the infinite coldness or the infinite darkness cube physically attracts the infinite matter, the infinite energy or the infinite light cube, which has density or is solid or is physical, and a physical part of the infinite matter, the infinite energy or the infinite light cube is broken off or is released, and takes up space in the magnetic sphere or takes up space in the magnetic circle formed in the universe or the vacuum of space by the two infinite magnetic cubes, [(((((((-) (1) (+))))))))].

Once the solid or the solid part or the solid piece of matter which is in the physical form of a smaller cube of matter, or simply call it the universe's solid egg, from the infinite matter, the infinite energy or the infinite light cube, physically enters the vacuum or the compression of space, there is the actual or the real and the physical beginning or the physical laying of the universe's egg.

Since empty space is super cold and also a super or a giant vacuum, the physical properties of matter, energy or light, or the smaller matter, energy or light cube as an egg in the vacuum of space, physically changes.

Matter, energy or light now in the super vacuum or in the super compression of space shrinks or matter, energy or light is super compressed in the super vacuum of space.

Once super compressed, matter, energy or light as a smaller cube is now super compressed into a sphere of matter, energy or light.

The new sphere of matter, energy or light in the super vacuum of space super heats and super explodes and thus simply expands throughout the universe or simply expands throughout the simple magnetic sphere or simply expands throughout the magnetic circle or magnetic field or magnetic shield.

Thus, the Big Bang!

The Universe / Forester de Santos

Actually, the Big Bangs, all 118 of them!

Chapter Two

[((((((((-) (1) (+))))))))]

As it can really be seen

As it can be really be seen, the rest of the universe or the rest of actual or real existence or even creation is time, physical evolution, history and simple ideas, true or not.

That is to simply say, the rest of the real or the actual universe or the rest of the real or actual reality or physical existence or physical creation is simply the Big Bangs of human ideas!

The reality of the matter is that there are really two creations.

There is physical creation or physical reality.

And there is mental creation, but mental creation or human thought does not let physical creation or actual reality thus be seen, and much less enjoyed.

Just because it is believed, for example, it does not make it so.

The truth or reality is not something one just believes in, but the truth or reality is something that can really be seen.

When mankind does not understand something or even see it, mankind humanizes it or makes it unnatural.

The universe or existence or physical creation is as the conscious living being simply is.

The more the conscious living being simply is or the more the conscious living being simply becomes or really knows or understands, the more the universe or the more physical or the more actual existence or the more physical reality or the more physical creation simply will be!

In other words, the universe or existence herself or even physical creation is only limited to the conscious thought or only limited to the conscious belief of the conscious living being.

Limit the conscious knowledge of the universe or limit the knowledge of actual existence or of actual physical reality or of physical creation, thus limit the conscious living being.

Existence, the universe or physical creation is composed or really made up of three main parts.

The three main parts are physical reality, neutral reality or point of view, and the reality not seen but can be negative or irrational.

Or simply put, all of Existence is "(+) (0) (-)."

In other words, $E = +0-$. That is, plus zero minus.

The same formula can also be used for the universe.

That is to really say, "$U = (+0-)$ or $(-0+)$."

Thus, the universe is really equal to the actual reality recognized plus personal point of view or personal understanding or the reality actually seen minus the reality not yet recognized or the reality believed.

The greater the plus is the greater the point of view, and the lesser the negative or the lesser the irrational.

The greater the negative or the more the irrational, the less the point of view or the less of actual physical creation is really seen and, therefore, the less understood and the less enjoyed or the less acknowledged.

But if the understanding or if the enjoyment or if the acknowledgement is greater or taller, thus greater or taller is the universe or thus greater or taller is existence.

Chapter Three

[(((((((((((U))))))))))))]

Dimensions of time

In reality, there is a finite or there are only 118 separate universes.

In reality, there is a finite or there are only 118 dimensions of time between the two infinite magnetic cubes: the space, the cold or the darkness cube, and the matter, the energy or the light cube.

Even though the two infinite cubes are encircled in magnetic spheres, the first 118 magnetic spheres are actual universes or are actual dimension of times.

The first 118 magnetic spheres are actually the middle, the center or the belly bottom of existence or of reality or physical creation.

Every magnetic sphere, for example, illustrates a separate universe or a separate dimension of time, [(((((((0))))))))].

The number of universes or the number of dimensions of time, once again, is in reality finite or the number is limit even though the magnetic spheres are countless as the numbers are infinite or countless.

What this really means, is a magnetic sphere inside a magnetic sphere inside a magnetic sphere until infinity, [(((((((0)))))))]...

However, only the first 118 magnetic spheres are actual universes or are actual dimensions of time.

Each magnetic sphere or every universe is separated by a dimension of time or is separated by a dimension of space.

The magnetic circles or the magnetic spheres increase in size as the two infinite magnetic cubes extend forever or extend until infinity.

The motion, the movement or the expansion in Time or in the universe or the expansion of the universes is created or is caused by the magnetic field of the two infinite magnetic cubes.

By the way, not every magnetic sphere is a dimension of time and not every magnetic sphere is a separate universe.

The dimensions of time are finite as the universes are finite in number. There are only 118 dimensions of time as there are also only 118 separate universes.

When empty space or the vacuum of space attracts the infinite matter, the infinite energy or the infinite light cube, not only does the infinite matter, the infinite energy or the infinite light cube releases a smaller matter, energy or light cube with 118 smaller matter, energy or light cubes, but also the infinite matter, the infinite energy or the infinite light cube releases multiple matter, energy or light cubes at the very same time.

The infinite matter, the infinite energy or the infinite light cube releases smaller matter, energy or light cubes with 118 smaller cubes each according to the size of the magnetic spheres or the size of the universes.

That is to say, the infinite matter, the infinite energy or the infinite light cube releases at the very same time one matter, one energy or one light cube with 118 pieces of matter, energy or light for the number one sphere or the number one or the very first universe; the infinite matter; the infinite energy or the infinite light cube releases two matter, energy or light cubes with 236 pieces of matter, energy or light for the number two or the second sphere or the number two or the second universe; the infinite matter, the infinite energy or the infinite light cube releases three matter, energy or light cubes with 354 pieces of matter, energy of light for the number three or the third sphere or releases three matter, energy or light cubes for the number three or the third universe, and so on until the very last universe or on until the very last dimension of time, which is dimension number 118. Dimension number 118 or universe number 118 is 118 times the size of dimension number one or 118 times the size of universe number one.

In other words, magnetic sphere number118 is 118 times the size of magnetic sphere number one. Universe number 118 is really 118 times the size of universe number one.

For every one cube of pure matter, energy or light, there are 118 smaller matter, energy or light cubes or 118 pieces of matter making that one matter, energy or light cube.

The very same number of matter, energy or light that enters the vacuum of space is the very same number of space taken or occupied by the matter, the energy or the light.

As soon as matter, energy or light from the infinite matter, the infinite energy or the infinite light cube enters empty space, cold or darkness or matter enters the vacuum of space or enters the magnetic spheres created by the magnetic field of both the infinite matter cube and the infinite space cube, matter, energy or light in the vacuum of space is super compressed due to the enormous pressure or due to the enormous vacuum of space, cold or darkness.

The enormous pressure or the enormous vacuum in space causes matter, energy or light in the vacuum of space to super compress into a giant or into an enormous magnetic sphere of matter, energy or light in the vacuum of space and to superheat and to super explode, thus making possible the Big Bang!

Actually, making possible the Big Bangs! Or the hatching or the hatchings!

Enormous energy is loss from the super Big Bangs! Or from the super hatchings! But a lot of matter, energy or light is still left in the vacuum of space to spread or to expand throughout the vacuum of space and spread or expand throughout the universes or throughout the magnetic spheres.

Some matter will lose energy, therefore, becoming stars, planets and moons and some matter actually becoming even smaller pieces or bits of matter, such as space rocks or space dust.

The heaviest or the denser matter or the heaviest or the denser stars at one point in time or at one point in the vacuum of space will turn into supernovas and then implode into super black holes or turn into the cancer of the universe and universes.

Actually, black holes or collapsed stars are like giant vacuum cleaners or giant space sweepers, really giant magnetic graves which will destroy all matter, energy or light!

There will be another point in time or another point in the vacuum of space when most of the universe and universes will be occupied by super or by giant black holes or by giant cancerous stars or dark stars or vast empty magnetic graves.

Once again, the universe or universes will be occupied by giant vacuum cleaners or giant sweepers or giant graves.

The universe and universes will look like a giant Swiss cheese, but without the appetizing smell.

At that very stage, when the entire universe and all universes are occupied by enormous black holes or enormous dark stars or enormous vacuum clears, the enormous black holes or the enormous dark stars will disintegrate or will drain the entire light or the entire energy or the entire matter of the universe and universes spontaneously or at once!

That is to say, the giant black holes or the giant space vacuum clears will begin contracting or moving or going to the center of the vacuum of space or to the actual physical point of impact or to point zero or to the point of origin until the entire physical universe and the other universes collapse—the exact opposite of the Big Bang! Or Big Bangs!

The black holes or the collapsed stars will begin to move toward the center or begin moving toward the actual physical beginning or the physical origin of the universe and universes, removing all the matter, all the energy or all the light from the entire universe and universes and thus making one giant or making one enormous black hole or one giant or one enormous vacuum per universe or one enormous vacuum per magnetic sphere.

It is this new giant or this enormous black hole vacuum in the vacuum of space or inside the magnetic sphere that actually attracts new matter, new energy or new light from the infinite matter, the infinite energy or the infinite light cube.

Once new matter, new energy or new light from the infinite matter, the infinite energy or the infinite light cube enters or re-enters the giant black hole vacuum in the vacuum of space or in the magnetic sphere, the new matter, the new energy or the new light is super compressed recreating the Big Bang! Or really creating the Big Bangs! And destroying the giant or the enormous black hole vacuum!

Actually, every giant or every enormous black hole in the vacuum of space or in every occupied magnetic sphere or in every occupied dimension of time or in every occupied dimension of space will be destroyed at once or at the very same time by new matter from the infinite matter, the infinite energy or the infinite light cube!

In other words, every giant black hole vacuum occupying the 118 universes or occupying the 118 dimensions of time will be destroyed at the very same time by new matter, new energy or new light from the infinite matter, the infinite energy or the infinite light cube.

However, there also exists the possibility that the giant black hole or the black holes will disperse as if a storm and space will become as if a sea of tranquility.

Chapter Four

The Stars and the Black Holes

The universe is a neutral point or is a point zero which is converted into a positive point or into a one or more when light or matter enters into the universe.

The universe also is converted into a negative point when the grand majority of light turns off or the majority of matter no longer has energy…

Before light or matter enters into the universe the weight of the universe is cero and when light or matter enters into the universe thus the universe takes on weight.

Light or matter is composed of 118 Elements and the weight of each element is two times its atomic number, more like its positive number.

In the case of element number one, for example, its weight is of two and in the case of element number two its weight is of four and in the case of element number 118 its weight is of 236…

Curiously, that the totality of the 118 elements adds to one and that the totality of their weight adds to 2.

That is to really say, that if we added from one to 118 thus the sum would be 7,021.

And if we added that sum thus it would be 10 and if we add that last sum thus it would be one. That is to really say, 118 is equal to one...

And if we added from two to 236 the sum would be 14, 042.

And if we added that sum thus it would be 11 and if we added that last sum thus it would be 2. That is to really say, 236 are equal to two...

Thus, light or matter enters into the vacuum of space or into the universe as one or as a unit which is composed of 118 pieces or elements and the weight of the element is two times the atomic number of the element.

Thus in truth, number one itself is composed of 100 percent plus 18!

That is to say, that the number one or even oneself is equal to 118 percent!

Also light or matter could enter into the vacuum of space or into the universe with only or as one element with its weight of two but that element would be able to convert into the other elements, even to the element 118 and its double weight of 236...

Element number one, for example, enters into the vacuum of space with its weight of two and it has 117 other possibilities of converting into the other 117 elements.

That is to say, element one is composed of 118 parts or pieces or the 118 percent and element number has 117 possibilities of converting into the other 117 elements according to the weight which element number one maintains.

In the same manner, element number two with its weight of four has 116 possibilities of converting into the other 116 elements or until the element number 118 with its weight of 236...

When light or an element enters into the vacuum of space, light or the element enters as if it were a piece of magnet or as if it were a bar magnet.

In the vacuum of space the magnet or light or the element is super compressed not only until it takes the form of a sphere or round but also light or the element or the magnet is super compressed until it gets to a very high level of temperature.

And when the temperature gets to its highest level thus light or the magnet or the element super explodes causing the light or the magnet or the element to divide into two parts and the part with less weight, such as the negative part, takes position in the magnetic field and that magnetic field now is a negative field...

The superior or the positive or the heaviest part of light or of the magnet or of the element takes position in the center or in the nucleus.

Thus, now we have the negative part of the magnet going around the positive part when before they were united and the center or the nucleus was neutral...

And even though the element lost weight because of the explosion or because of bursting, the element continues the one for two.

That is, its weight continues of two although the element one now is 0.999 and its weight is double, 1.998...

Thus, now element number one was reduced to about 0.999 with its new weight of 1.998 but to element number one also remains a neutron or even more than one or a neutral part which can be converted or can be transformed into a positive part and that way not only adding the number of the element but also adding its weight even though it will have only one negative particle going around the center or the nucleus or the positive side...

But if one or light or the magnet or the element does not convert into the next number or into the next element thus it loses its energy and will only be a piece of dead matter in the vacuum of space and it will be removed one day by the black holes...

Thus, light or the element is the same as a negative particle, is the same as a neutral particle and is the same as a positive particle.

In a way, one is equal to a negative portion plus a neutral portion plus a positive portion which totality is of 0.999 after entering the vacuum of space.

But in the vacuum of space light or the element is positive even though the vacuum of space is neutral but reacts as if negative because of the vacuum.

And when the element increases its positive part by converting the neutral part into a positive part, the element cannot attract the negative part because of the vacuum of space because now the negative part becomes as if more or its weight increases because of the weight that it receives indirectly from the vacuum of space.

And if there were not positive attracting the negative thus the negative would expand through the vacuum of space and it would stop being negative and it would be dead matter...

Now, an element, in this case a star, which number is high as the same as its weight thus lasts or remains longer in the vacuum of space or in the universe.

But the element or the star becomes heavier while the energy or matter to continue on lasts or it begins to transform from positive to neutral and once neutral, the element or the star practically becomes negative when its excess or super weight attracts the electrons toward the center and thus causing an implosion in where the element or the star becomes a nova or a new star but without light or without energy and that way causing an enormous hole in the vacuum of space when before the element or the star occupied the vacuum of space as an element or as a star...

In other words, matter or the star in the vacuum of space changes from positive to neutral and then from neutral to negative.

Thus, + 0 -, in where the negative is a black hole or a super vacuum cleaner in the vacuum of space.

This black hole practically eats or sucks all the matter around it to take all matter out from the vacuum of space or from the universe to make new space for new matter or for another beginning...

But as long as there are black holes in the vacuum of space or in the universe, thus the vacuum of space or the universe is negative and as long as the vacuum of space or the universe continues as or is negative thus it keeps being for something and not for neutral and from neutral or from zero to positive...

Thus, so that the universe becomes neutral or to zero and from neutral or from zero to positive thus all the black holes or super space vacuums must stop from functioning and once the black holes or the super space vacuums stop from functioning for lack of matter or for lack of energy thus the vacuum of space or the universe will return to neutral or to zero…

Now then, this new vacuum in space or in the universe is neutral and has zero energy but even so attracts new matter or new stars or attracts the new or the next beginning which is outside of the universe or outside the vacuum of space…

And once new matter or new stars enter into the vacuum of space or into the universe thus the vacuum or the universe will be positive or one or more…

Thus, the cycle or the model or the rhythm of (+ 0 -) continues until the end of all the times…

Chapter Five

The Universe and the Number

The Universe as is Creation truly is about giving name or naming to start or to begin and giving rename or renaming to continue on without beginning again even though to give rename or renaming is to become as if forever new and as if never there were a start or a beginning and neither an end…

The Universe as is all of existence truly is composed of numbers or be it positive numbers or be it neutral numbers or be it negative numbers or be it a combination of all the numbers at the very same time.

But there will always be a number that is greater than the other negative numbers until that positive number is converted into a neutral number and after that positive number is converted into a neutral number thus also it will be converted with time into a negative number or a void.

In other words, the Universes as is existence really is equal to (+ 0 -), but existence does not return to negative but rather a part of her and that part of her is the universe or matter or light, which returns to or truly is converted from positive to neutral or to zero and much later it is converted to negative or into nothing.

The number zero is not negative or nothing or void. The number zero only is a neutral number which could be converted into a positive or negative number.

Existence as matter or as light or energy is truly composed of three main parts which add to 118 percent, 118 times 3 because they truly are 118 positive parts, 118 neutral parts and 118 negative parts or (+ 0 -).

Now then, the neutral part or part zero is the universe and it is where it is added, it is subtracted, it is multiplied and it is divided at the very same time.

The positive part is the physical part of existence or the part from where comes out matter or light or the elements which truly are pure matter, but this part of existence appear to be minor than the other parts of existence.

The negative part of existence is the lack or is the vacuum of space or darkness which is composed of three parts, one is space or emptiness and the other two are space-time, which truly are created by the interaction of the magnetic field between the positive and the negative.

When matter from the physical side of existence enters into the vacuum of space, matter even though a single piece, matter enters with its 118 parts assimilating the 118 elements.

That is to say, if only one element enters into universe or the vacuum of space that element could assimilate or even can come to be converted into the other 117 elements…

But in reality matter enters from the physical side of existence into the universe or the vacuum of space in 118 pieces into 118 parts of the universe or the vacuum of space or into 118 dimensions at the very same time.

That is to say, into the universe or into the vacuum of space there enters 118 elements and every element has 118 parts and at the very same time there enters 118 elements into 118 dimensions.

But that, however, does not remain like that, because with every element the 118 parts are also multiplied.

In other words, element number one has 118 parts but element number two has 236 parts and element number three has 354 parts and this pattern continues on until the last element, which is element number118.

Also, the weight of the element is twice its atomic number. That is to say, element number one has a weight of two and element number two has a weight of four and element number three has a weight of six. This pattern also continues on until element number 118.

Interestingly, however, that all the parts of an element add to the atomic number of that element!

Thus, element number one has 118 parts and if we added those 118 parts thus we would get one. That is, if we added 1+1+8 thus they would give us10 and if we added 1+0 thus we would get 1.

In the case of element number two, if we added 2+3+6 thus we would get 11 and if we add 1+1 we would get 2. This pattern continues on until element number 118. Element number 118 has 118 parts and every part also has 118 parts thus giving a sum of 13,924, also adding to one.

Now then, when element number one enters into the vacuum of space element number one enters with one proton, with one neutron and with one electron, (1+0-1). And also element number one enters with a weight of two.

But once in the vacuum of space, space comprises the element and the element is comprised causing the element to super heat and thus also causing the element to burst or explode.

In that burst or explosion the element loses a neutron or a third part and no longer the weight is of two but is less, such as 1.67.

Also the electron was separated and now is spinning around the proton or the nucleus of the element, (+) -.

In the case in where an element has a high number thus that element will have neutrons in its nucleus, such as (+0+) -, -.

Or (+0+0+) -, -, -. This new transformation of the element, in where the electron is separated from the nucleus makes it possible for the element to be united to other elements and that way converting into a mixture of elements or isotopes...

Furthermore, before an element enters into the vacuum of space its three main parts have the same size or the same weight, ((+) (0) (-)).

But once that element enters into the vacuum of space thus that elements is divided into two parts, the nucleus in where now is the proton (+) and the neutron (0), and the outside part in where now is found the electron (-) going around the nucleus, ((+) (0)) (-).

And while the nucleus keeps its size or weight, the electron loses its size or weight because of the interaction or friction which it has with the nucleus or with the center of the element.

The interaction or friction which the electron has with the nucleus or with the center of the element also causes the electron to last less or lasts less time in the vacuum of space.

Once the electron is fused or is exhausted, the element or the nucleus is converted into a neutral element or without energy even though the nucleus is still positive or with protons and neutrons.

But the outside part or the electric field of the element is now a neutral field or the electrons have been converted into neutrons because of lack energy.

In other words, the seven electron rings of the element or the atom now are neutral when before they were negative because of the electron.

And just as the element functions thus that way also functions the number and existence herself, but the number or the symbol of the number is only an illustration of the numbers but truly does not show how is the number in existence or outside the universe or the vacuum of space in where there is no friction or movement even though there is a magnetic field.

Thus in truth, the number one or 1 outside the vacuum of space is represented by a cube.

Now, the cube or the number or the element one is composed of three main parts and they are the positive part, the neutral part and the negative part, (+) (0) (-).

But those parts also are composed of 118 other parts. That is to say, that the positive part also is composed of 118 parts and the neutral part is also composed of 118 parts as the same as the negative parts which also are composed of 118 parts.

And when the cube or the number or the element enters into the vacuum of space the cube or the number or the element is compressed into a sphere or into a globe but still with its 118 parts.

Thus, the number one is composed of not only 100 percent but also of another 18 parts or of another 18 percent. And if we added the parts thus we would have 1.

Now then, the number one or the symbol 1 as also is all of creation is a continuation because the start or the beginning is zero or a point or a neutral or an empty space.

But the number one or the symbol 1 also represents all of existence, the physical part, the neutral part and also the negative part. Also the number one has the ability of converting itself into its 117 other parts also with their other 118 parts.

In other words, the number or element one also has the ability of being infinite because also it could renew into a greater number such as the number two.

And that makes it possible the other 117 parts which add to 9 and the number 9 is a symbol of renovation.

That first renovation extends the time of the number or of element one.

Thus, if we added the 117 parts which remain to one plus its other two parts, the neutral with its 118 parts, and the negative with its 118 parts, the sum would be of 353.

And if we added 353 thus it would give us 11 or eleven and if we added 11 we would get two, the possibility or the ability of the number or element one if it is renewed.

And once that the number or that element number one has renewed as two, thus the number or element one has become or will continue as double or for much more as two and as double the abundance.

And the very same step or process is with the number or with element number two. If we added all the parts that the number or the element two has, which are the double of one, thus it would give us four or 4.

That is to say, if we added all the parts of the number or element two thus we would get the double.

Conclusion

The universe is finite but existence is infinite or without size.

The universe ends or stops running but existence runs or she expands toward all sides at the very same time and at the very same time existence also compresses toward all sides, that way also adding to her very self an infinite weight and an infinite size.

That is to say, existence adds to herself, she multiplies herself, she divides herself and existence also subtracts her very self or from her very self to be able to become forever for more and even for new.

The universe begins but existence forever existed and existence forever will exist. Existence has no time even though time existence is.

Existence is also composed of three principal parts which all add to one and that one is the grandiose part which makes the difference in all of existence.

Existence consists of matter, that which is physical. Existence also consists of space or of vacuum, that which is lack of something or lack of matter, energy or light.

And existence also consists of movement or of that which many call space-time, which is a movement as if in waves in space.

Thus in truth, existence is a simple magnet!

In other words, existence really is composed of +0-; in where (+) is equal to positive; in where (0) is equal to neutral; and in where (-) is equal to negative or the lack of.

In the scale of colors would be white, grey and black; in where white is light or matter; grey is neutral or space-time; and black is darkness or lack.

Now then, just as existence truly is composed of three parts and practically of parts opposite to the other parts, +0-, thus one also truly is composed of those very same opposite parts but in knowledge because one is knowledge.

And one can truly be positive knowledge or be neutral knowledge or even be negative knowledge.

But the greater the positive knowledge one has, less the neutral knowledge and less the negative knowledge or less the lack of negative knowledge.

In other words, the greater the knowledge of reality that one has, the greater the reality of one and less is the reality not known or less is emptiness or darkness.

Thus in truth, the more the knowledge of one, the more the abundance or the more is the life of one or more life makes sense to one.

Interestingly, that existence functions the same as one but existence knows it not but existence will never stop from being or will never stop from existing for all the eternity of eternity because she truly makes herself for much more!

That is in truth, even though all of existence is one, because of one existence truly is and for all eternity infinite!

And every time that existence is increased by one or is increased by more than one, the point of neutrality or the neutral is less as the same as the negative or the unknown point is less.

But making the things less does not make existence greater, but making the things or existence greater makes existence greater!

And just as existence makes the things greater for existence to be greater herself, oneself can be greater by recognizing that life is more or that in life herself there is more and that there is with all peace, the neutral, and that here is with all knowledge, one or more than one; and also that there is with all gladness and with all joy and that also there is with all abundance of life, life renewed…

Thus, through knowledge one renews for much more and because of one renewing for much more, one truly can continue for all of eternity as the very eternity…

Further Notes

Just as the black hole removes all dead matter from the vacuum of space or from the universe to make the vacuum of space or the universe neutral to attract need matter or light or illumination from outside the vacuum of space or outside the universe, thus we should do the same with our conscious mind with our dead beliefs or with our dead faith which will take us dead to the grave if we do not remove our dead beliefs or our dead faith and that includes blind faith from our conscious mind...

Seek the Truth

Seek of the truth or seek the truth and when she let herself be found from you, you truly will become taller, bright and will even have more confidence in yourself.

The very curious thing about the truth is that she begins with one and if one truly does the movement to find her, thus she will come to one or she will be given to you by he that made or that presented her for the truth.

A Higher Mental Consciousness

The path of less resistance or without contention not only leads to laziness or to vanity or to nothing but also the path of less resistance or without contention leads to death.

And death ends any possibility to any other possibilities, such as a higher mental consciousness or illumination with real gladness and with real joy, in double abundance all five portions, and also with the power and with the authority of the heavens here on the earth.

The (0+-) Factor

Those that accept an idea or ideology blindly or without trying or testing that idea or ideology, thus they unknowingly become liars.

But nonetheless, liars they are and as liars they live and will lie to others to convince them to accept or believe and they will keep on lying until death herself and death herself as a black hole will close their lying trap.

Now, when one does the movement or gets interested for knowledge, thus one truly becomes that knowledge, true or not.

When the universe or creation came to exist, the universe or creation came to exist because of knowledge, but true knowledge even though most is unseen.

The proof is in the numbers or in matter which is composed of elements and the elements in turn are really numbers, real numbers.

In other words, matter is knowledge because matter is composed of elements and the elements are composed of numbers and the numbers are composed of positive, neutral and negative states, thus the, (+ 0 -), plus zero negative factor or a magnet with its three opposites parts.

What this really means is that knowledge can become neutral or void or useless and then become negative or contra productive if at first nothing was done with that knowledge, such as using or converting knowledge into acknowledgement or illumination or a positive or useful respond.

In the same manner as above, the universe or the vacuum of space becomes positive when matter or light enters the universe or the vacuum of space.

But as matter or light begins to burn out, the universe or the vacuum of space also begins to turn neutral until it turns negative, negative because of the black holes which now rule the universe or the vacuum of space which they now also suck up any remaining dust or matter or light to make space for another beginning.

But this new beginning is as if the very first beginning because there will not be any trace that there was ever a first beginning.

But the above matter or knowledge or illumination would only be a simple theory or an idea or ideology if the conscious being, which is also real knowledge, did for real acknowledgement or illumination and the conscious being would really have the power and the authority over matter or light to refresh matter or light and thus keep the universe or the vacuum of space always or for all eternity positive and refreshing in double abundance, all five portions of her!

Who am I really?

My pen name as a writer is Forester de Santos and I am truly on a very grandiose crusade of rebirth alive or to be born again into royal life, which really is a higher mental consciousness, with complete gladness and with complete joy and even also with complete abundance of God but as much more than God and as much more than Creator.

God equals true or real knowledge or true or real acknowledgement, which can only be achieved through a higher or taller mental consciousness or awareness...

Now then, one who truly is on a very grandiose crusade cannot follow another or he cannot let himself be surrounded by his beloved ones or his fans because he cannot cross over them or he cannot cross over because of them being in the way or because of them blocking the path to righteousness which really is but which cannot be seen until rebirth or until one is born again or until one enters a higher or taller mental consciousness, which really is an expansion of mental consciousness or self-awareness with complete gladness and with complete joy and also with complete abundance.

I do not, therefore, ask to be followed, not because I will not lead, but because I will not look back, because I am not going that way, but as I move forward and ahead I will only look to my right and to my left to see who walks with me.

But so far only one walks besides me, at times in front of me and often on my left side…

But those that truly decide to follow me will become as me and as me will truly receive or truly gather true knowledge because my struggle or my contention or my very grandiose crusade of rebirth is true, so true in fact that I have become a much better person because of the true faith which I have come to receive through my search and research for the truth.

And because I have come to have true faith or faith of God, which is true knowledge, thus I use my true faith as a shield to repel or to reject other beliefs or good sounding lies that will only take one dead to the grave!

Therefore, to rebirth alive or to be born again, really a higher mental consciousness, while still living here on the very earth which will be as in the very heavens through rebirth!

Now then, rebirth or life renewed is really about a higher mental consciousness!

The Universe / Forester de Santos

$$[(\text{-} (\text{-} (\text{-} (\text{-} (\text{-} (\text{-} (0 + 1) +) +) +) +) +) +)]$$

If you truly enjoyed this simple and humble work, please leave a comment according to your good pleasure and give also a rating but also according to your good pleasure.

Thanks so very much for your time and best of wishes, Forester de Santos.

Thanks for reading my work!

0+1 = peace and knowledge to all mankind...

www.ingramcontent.com/pod-product-compliance
Lightning Source LLC
Chambersburg PA
CBHW030733180526
45157CB00008BA/3152